United States Nuclear Regulatory Commission

Protecting People and the Environment

Public Information Circular For Shipments of Irradiated Reactor Fuel

Office of Nuclear Security and Incident Response

AVAILABILITY OF REFERENCE MATERIALS
IN NRC PUBLICATIONS

United States Nuclear Regulatory Commission

Protecting People and the Environment

NUREG-0725, Rev. 15

Public Information Circular For Shipments of Irradiated Reactor Fuel

Manuscript Completed: January 2010
Date Published: May 2010

Prepared by
A.G. Garrett, S.L. Garrett, and R.G. Ostler

Pacific Northwest National Laboratory
P.O. Box 999
Richland, WA 99352

K. Jamgochian, NRC Project Manager

NRC Job Code N4104

Office of Nuclear Security and Incident Response

ABSTRACT

This report provides information on the shipment of irradiated reactor fuel (spent fuel) subject to regulation by the U.S. Nuclear Regulatory Commission (NRC). It briefly describes spent fuel shipment safety and safeguards requirements of general interest, summarizes data for highway and railway shipments from 1979 - 2007, and lists, by State, recent highway and railway shipment routes. This circular does not include Department of Defense and Department of Energy spent fuel shipments.

The enclosed route information reflects specific NRC approvals that the agency has granted in response to requests for shipments of spent fuel. This publication does not constitute authority for carriers or other persons to use the routes to ship spent fuel, other categories of nuclear waste, or other materials.

PAPERWORK REDUCTION ACT STATEMENT

CONTENTS

FIGURES

TABLES

1.   INTRODUCTION

The Atomic Energy Act of 1954, as amended, authorizes the U.S. Nuclear Regulatory Commission (NRC) to regulate the commercial nuclear industry for the purpose of protecting the public health and safety and the environment from the effects of radiation from nuclear reactors, materials and waste facilities. We also regulate these nuclear materials and facilities to promote the common defense and security of the United States. Included in this authority is the regulation of certain aspects of the transportation of irradiated reactor fuel (spent fuel). Section 2 of this report explains the NRC's role in this regard. Section 3 provides descriptive statistics on spent fuel shipments in the commercial nuclear industry for the period of 1979 - 2007. Section 4 lists highway and railway segments used in each State for transporting spent fuel in recent years (1998 - 2007).

2. REGULATORY REQUIREMENTS FOR SPENT FUEL SHIPMENTS

The NRC regulates spent fuel shipments in terms of both safety and safeguards. Safety involves the protection of public health and safety during transport as well as in the event of handling or transportation accidents, whereas safeguards relates to the protection of shipments against deliberate, malevolent acts by persons.

2.1 Safety Requirements

The NRC and the U.S. Department of Transportation (DOT) share Federal regulatory responsibility for spent fuel transportation safety. In terms of safety, the NRC ensures that spent fuel packaging meets strict regulatory design rules, approves package designs and quality assurance programs, and conducts inspections. Title 10, Part 71 "Packaging and Transportation of Radioactive Materials," of the *Code of Federal Regulations* (10 CFR Part 71) specifies NRC packaging requirements. The role of DOT in regulating spent fuel shipment safety is broad and covers all aspects of actual transportation, including route selection, vehicle condition and placarding, driver training, packaging marking, labeling, and other shipping documentation, as specified in 49 CFR Parts 171-178, 356 and 397.

2.1.1 Packaging Standards

The NRC regulations for spent fuel shipments primarily address the packaging of spent fuel to protect public health and safety during transportation. The packaging standards that have been established in the regulations provide that a spent fuel package (cask) shall prevent the loss or dispersion of the radioactive contents, provide adequate shielding and heat dissipation, and prevent nuclear criticality under both normal and accident conditions of transportation. The regulations specify the normal conditions of transportation that must be considered in terms of hot and cold environments, pressure differential, vibration, water spray, impact, puncture, and compression tests. Accident conditions that must be considered are specified in terms of impact, puncture, fire conditions, and immersion.

The NRC reviews the cask design to verify its resistance to accident conditions. The NRC must issue a certificate before a cask fabricated from the reviewed design can be used to transport spent fuel.

The ability of packaging to provide protection has been demonstrated by the responses of packaging during actual traffic accidents. For example, an accident occurred on December 8, 1971, on a major highway near Oak Ridge, Tennessee. In this accident, the driver of a vehicle carrying a spent fuel cask swerved to avoid colliding with an oncoming vehicle, lost control, and overturned off the roadway. The cask assembly was thrown into a ditch, traveling more than 100 feet before coming to rest. No release of contents or radiation occurred. The outer surface of the cask sustained minor damage. The spent fuel cask was placed on another trailer and taken to its destination. The cask was returned to service after inspection and repair of the minor damage.

Controlled tests have also demonstrated the resistance of casks to accidents. In one test, sponsored by the U.S. Department of Energy (DOE), a truck bearing a cask was deliberately placed in the path of and was struck by a 120-ton locomotive traveling about 80 miles per hour. In another DOE test, a cask aboard a truck moving about 80 miles per hour was deliberately crashed into an immovable concrete structure. Subsequent examination in both these tests confirmed that the casks would not have released any radioactive material had they been loaded with spent fuel. Thus, both field experience and controlled tests have substantiated the strategy of depending on packaging design for safety in transit.

Further information on spent fuel shipments safety and cask performance-oriented requirements is available in NUREG/BR-0111, "Transporting Spent Fuel - Protection Provided against Severe Highway and Railroad Accidents" and NUREG/BR-0292, "Safety of Spent Fuel Transportation."

2.1.2 Routing Requirements

An NRC-approved route must meet DOT requirements in 49 CFR 397.101, which provide that radioactive material shipments are to be carried over the Interstate System of highways and available city beltways as the primary roadways. No routing rule exists for rail shipments. Roadways are designed as preferred highways based on their capacity for reducing transit times. Appropriate State agencies, following prescribed criteria, may designate an alternate route to the preferred Interstate System. Spent fuel carriers are responsible for abiding by the routing rule when they transport spent fuel by highway.

2.1.3 Spent Fuel Shipment Safety Record

The safety record for spent fuel shipments in the United States and in other industrialized nations is enviable. Of the thousands of shipments completed over the last 30 years, none has resulted in an identifiable injury through the release of radioactive material.

2.2 Safeguards Requirements

In May 1979, the NRC revised its regulations in 10 CFR 73.37, "Requirements for Physical Protection of Irradiated Reactor Fuel in Transit," to strengthen the protection of shipments of spent fuel against radiological sabotage. In 10 CFR 73.37(a), the regulation identifies the material requiring physical protection as "... a quantity of irradiated fuel in excess of 100 grams in net weight of irradiated fuel, exclusive of cladding or other structural or packaging material which has a total external dose rate in excess of 100 rems per hour at a distance of 3 feet from any accessible surface without intervening shielding...." The NRC subsequently revised these regulations in May 1980, in response to public comments. The regulations require, among other actions, NRC approval of routes for the transportation of spent fuel to ensure adequate planning for physical protection against actual or attempted acts of radiological sabotage. Physical protection requirements for NRC licensees who transport or deliver spent fuel to a carrier for transport include advance notification to the NRC of shipments, procedures for handling emergencies, establishing a communications center, contact with the communications center every 2 hours, a written log of shipment events, arrangements with local law enforcement agencies, avoidance of intermediate stops, surveillance of the shipment vehicle while stopped, armed escorts in heavily populated areas, escort training, onboard communications, immobilization devices on trucks, driver training, and notification of State governors before making shipments. In the aftermath of terrorist attacks on September 11, 2001, the NRC issued multiple safeguard advisories and security orders to enhance the security of spent fuel transportation. The advisories recommended that licensees implement additional security measures during shipments, including enhanced preplanning and coordination with the affected States, enhanced control and monitoring of shipments and enhanced emergency response procedures. The orders required specific additional security measures be implemented to ensure enhanced security when certain conditions were met.

2.2.1 Route Approval

NRC Licensees who transport or deliver to a carrier for transport spent fuel are required to submit proposed routes for such shipments to the NRC for approval, from a safeguards perspective, before the use of a given route. For highway shipments, the licensee must propose a route that conforms to DOT routing rules. As part of an NRC security inspection, the NRC can survey proposed routes for communication reception, location of safe havens, and other considerations. The NRC may approve a route for a single shipment or a specified series of shipments. Once the NRC has approved a shipment series, the route may be used for all shipments without re-approval of the route for each shipment, provided that the NRC is notified in advance of each shipment. The NRC approves a route for a stated period and may renew the approval after that period. All approvals expire unless renewed. NRC approval authorizes only spent fuel shipments and does not include other categories of nuclear waste material. From time to time, the NRC may authorize alternate routes or detours, as circumstances dictate at the time of shipment. In addition, detours may be taken without prior approval, in response to unforeseen circumstances that arise during a shipment. The NRC has provided criteria for determining when and how such detours may be taken in published regulatory guidance (NUREG-0561, "Physical Protection of Irradiated Spent Fuel").

2.2.2 Notification of State Governors

The NRC requires its licensees to notify the Governor or the Governor's designee before transporting spent fuel within or through the State (10 CFR 73.37(f)). The notification must be made in writing and postmarked at least 7 days before transport, if mailed, or delivered at least 4 days before transport, if sent by messenger. The notification must include the following information:

- the name, address, and telephone number of the shipper, carrier, and receiver
- a description of the shipment, as specified by DOT
- a listing of the routes to be used within the State
- a statement that the NRC requires shipment schedule information (provided as an enclosure) to be protected from unauthorized disclosure

The enclosure to the notification provides the following:

- the estimated date and time of departure from the point of origin of the shipment
- the estimated date and time of entry into the Governor's State
- a statement that schedule information must be protected from unauthorized disclosure until at least 10 days after the shipment (or 10 days after the last shipment of a series) has entered or originated within the State.

The licensee must also notify the Governor of scheduling changes that differ by more than 6 hours from the furnished schedule. Subsequent distribution of the schedule information is at the Governor's discretion, but NRC regulations require all persons who receive the schedule information to protect it from unauthorized disclosure.

2.2.3 Spent Fuel Shipment Safeguards Report

Safeguards incidents for all spent fuel shipments are those that involve attempts at radiological sabotage of spent fuel, or purposeful acts that threaten or result in significant degradation of the safeguards system used to protect the shipment. The regulations require licensees to notify law enforcement authorities immediately upon the occurrence of discovery of a safeguard incident to initiate an appropriate response. In addition, licensees must promptly report safeguards incidents to the NRC by telephone, followed by a written report.

3. DESCRIPTIVE STATISTICS FOR HIGHWAY AND RAILWAY SPENT FUEL SHIPMENTS FROM 1979-2007

This section provides descriptive statistics on the shipments that have occurred since the NRC began approving spent fuel shipments in 1979, through 2007. It includes only shipments of academic, industrial, and utility irradiated reactor fuel subject to NRC regulation. The NRC does not regulate DOE shipments; therefore, the statistics reported here do not include them.

Figure 3-1 is a graphic depiction of the approved spent fuel routes utilized between 1979-2007. Most of the routes depicted in the figure have expired.

Table 3-1 depicts the number of shipments and quantity of spent fuel shipped from 1979-2007.[1]

Table 3-2 and Figures 3-1 through 3-9 provide more detailed spent fuel shipment information, including routes used, mode of shipment (highway or railway) and shipment trends over time. Table 3-2 summarizes the spent fuel shipment data for the 1979-2007 period. For each year, the table provides four variables that describe shipping activity by mode. The quantity (i.e. weight) of material being shipped is based on the shipment pre-notification letters sent to the NRC in accordance with the requirements of 10 CFR 73.37.[2]

The NRC took the data for highway shipment miles for each route primarily from a road atlas and rounded the annual total to the nearest hundred miles. The kilogram-miles data are derived from shipment quantity and distance data and have been rounded to the nearest hundred thousand.

Table 3-3 shows the pattern of highway and rail shipments throughout the period 1979-2007. The number of domestic highway shipments (except for 1981) rose to a high of 209 in 1984 then declined until 1988, when the recent average of 9 highway shipments per year was reached. Import shipments have generally declined since 1980. The annual number of export shipments has been low (0-4) through the entire period. In addition, in 1990-1993 five international shipments passed in transit through U.S. ports.

Figure 3-2 shows that most (82.7 percent) of the 1554 spent fuel shipments during the 1979-2007 period were completed over highways. Figure 3-3 shows that most of the shipping activity occurred during the 1980-1987, with relatively low shipping activity after 1987.

Figure 3-4 shows that the larger quantity (84.4 percent) of spent fuel was shipped by railway, which reflects the greater capacity of rail spent fuel containers versus those for trucks. In

---

[1] Because of record keeping differences, the number of shipments and kilograms of spent fuel shipped in this table will be lower than the figures in Table 3-2.

[2] The NRC staff notes that the requirements of 10 CFR 73.37 do not result in uniform reporting standard for the quantity of radioactive material being shipped. Licensees providing advance notifications to the NRC use several different units to describe the quantity of radioactive material being shipped (e.g., "Pounds of hazardous material shipped." "Kilograms of uranium," and "kilograms of uranium-235 after irradiation"). For this report, the NRC staff estimated the quantity of material being shipped based on the information reported in the advance notification letter. For example, material testing reactor fuel assemblies typically weigh 4-6 kilograms (10-13 pounds) each. If the advance notification letter indicated that the shipment consisted of 10 of these assemblies, the NRC staff estimated the total weight of the spent fuel in the shipment as 60 kilograms (approximately 130 pounds).

addition, a few rail shipments included multiple spent fuel containers, further increasing the rail shipment payload. The figure indicates that 2,470,000 kilograms or about 2,400 metric tonnes, of spent fuel were shipped during the 1979-2007 period. Figure 3-5 shows that the greatest quantities of spent fuel were shipped during the period 1984-1987 and that since then, most spent fuel has been shipped by rail.

Figure 3-6 shows that the highway mode accounted for most (92.6 percent) of the 1,002,500 spent fuel shipment miles. Figure 3-7 shows that shipment mileage peaked in 1984, with a general decline in subsequent years.

Figure 3-8 shows the cumulative movement of spent fuel, calculated by summing the product of quantity and distance for all shipments and expressed in kilogram-miles. The unit is analogous to "ton-miles," a unit commonly used to measure the flow of commodities. The figure shows that the railway mode accounted for the majority (76.5 percent) of the 704.5 million kilogram-miles associated with spent fuel shipments. Figure 3-9 shows the kilogram-miles distribution by year.

Finally, Figures 3-10 and 3-11 provide an operational perspective for the spent fuel shipments. Figure 3-10 show the distribution of shipments by individual shipment quantity, and the corresponding total quantity shipped. The individual shipment quantities have been grouped into ranges, with highway shipments most frequently falling within the 0-10 and 400-500 kilogram ranges and with most railway shipments within the 6,000-16,000 kilogram range.

The smallest quantity range (0-10 kilograms) accounts for the largest number of highway shipments, 540. Conversely, the 214 railway shipments that exceeded 6,000 kilograms comprised 80 percent of the quantity of spent fuel shipped.

Figure 3-11 shows the distribution of shipments by distance range, and corresponding total quantity of spent fuel shipped. For highway shipments, the number of shipments generally decreases with shipment distance, although a significant number of shipments exceeded 900 miles. Of the 377,000 kilograms shipped by highway, 177,000 kilograms (47 percent) traveled less than 200 miles. The number of rail shipments was somewhat uniform over the ranges, with the 200-400 mile shipments accounting for 67 percent of the total quantity shipped by rail.

4. ROUTE SEGMENT LISTING FOR RECENT HIGHWAY AND RAILWAY SPENT FUEL SHIPMENTS

Table 4-1 lists highway and railway routes that have been used to transport spent fuel during 1998-2007. The table lists shipment origination and destinations, the approved route used, and the years during which the shipment was completed. The table shows that highway spent fuel shipments originated in 18 States, and that railway shipments originated in two States during the period discussed.

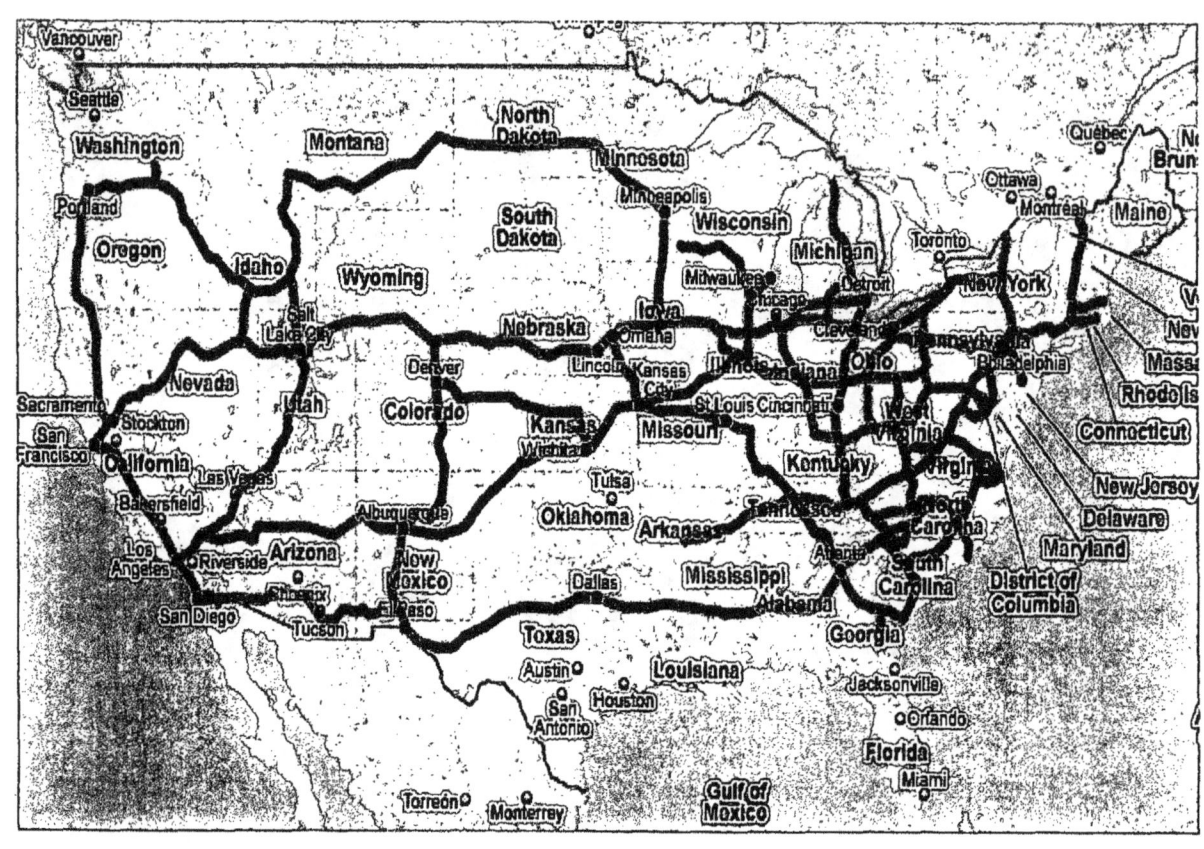

Figure 3.1 Routes Used for Spent Fuel Shipments: 1979 - 2007

Table 3-1 Number of Shipments and Quantity of Spent Fuel Shipped from 1979 - 2007

State	Origination	Destination	Number of Shipments	Kg of Spent Fuel Shipped
AL	Browns Ferry, AR	GE Vallecitos, CA	1	0
AR	Arkansas One, AR	Portsmouth, VA	1	3
CA	GE Vallecitos, CA	Argonne National Lab, IL	5	473.3
		Hanford, WA	3	102
		Idaho National Engineering Lab	18	71
		Port of Oakland, CA**	7	57
		Portland, OR**	1	1
		Richmond, CA**	3	39
	General Atomics, CA	Idaho National Engineering Lab	3	73
	Port of Oakland, CA*	GE Vallecitos, CA	9	93
	Richmond, CA*	GE Vallecitos, CA	1	12
	San Onofre, CA	GE Morris, IL	16	6800
	Univ of California (Berkeley), CA	Idaho National Engineering Lab	3	21
CN	McMaster Univ, CN	Savannah River Site, SC	1	5.6
CO	Fort St. Vrain, CO	General Atomics, CA	2	10
		Idaho National Engineering Lab	123	500
CT	Haddam Neck, CT	Battelle-Columbus, OH	3	1275
	Millstone, CT	GE Vallecitos, CA	3	43
FL	University of Florida	Savannah River Site, SC	8	1147
GA	E.I. Hatch, GA	Babcock & Wilcox, VA	1	20
		GE Vallecitos, CA	1	14
	Georgia Tech, GA	Savannah River Site, SC	1	4.47
	Port of Savannah, GA*	Savannah River Site, SC	5	65
IA	Duane Arnold, IA	GE Vallecitos, CA	3	25
ID	Idaho National Engineering Lab, ID	Portland, OR**	1	3
IL	Byron Station, IL	Alexandria Bay, NY	1	11

State	Origination	Destination	Number of Shipments	Kg of Spent Fuel Shipped
	Dresden Station, IL	Babcock & Wilcox, VA	3	56
		Portsmouth, VA	1	9
	GE Morris, IL	La Cross, WI	4	939
		Point Beach, WI	108	48382
	La Salle County State, IL	Portsmouth, VA	1	373
	Quad Cities, IL	Babcock & Wilcox, VA	2	499
		Battelle-Columbus, OH	1	18
		GE Vallecitos, CA	4	56
	Univ of Illinois, IL	Idaho National Engineering Lab, ID	2	745
		Univ of Texas at Austin, TX	1	73
	Zion, IL	Battelle-Columbus, OH	2	920
IN	Purdue University	Savannah River Site, SC	1	112
MA	MIT, MA	Savannah River Site, SC	20	629
	Univ of Mass - Lowell, MA	Savannah River Site, SC	2	194
MD	Calvert Cliffs, MD	Alexandria Bay, NY	1	25
		Battelle-Columbus, OH	3	64
	Dundalk, MD*	GE Vallecitos, CA	3	48.8
		Port of Oakland, CA**	1	0.302
	NIST, MD	Savannah River Site, SC	15	952.7
MI	Big Rock Point, MI	Portsmouth, VA	2	14
	Michigan State Univ., MI	Denver Federal Center, CO	1	8
		Idaho National Engineering Lab	2	11
	Univ of Michigan, MI	Savannah River Site, SC	24	986
MN	Monticello, MN	Battelle-Columbus, OH	2	67
		GE Morris, IL	29	195013
		GE Vallecitos, CA	4	55
MO	Callaway, MO	Alexandria Bay, NY	1	14
	Univ of Missouri, MO	Idaho National Engineering Lab	15	81
		Savannah River Site, SC	68	2380.3
NC	Brunswick, NC	Battelle-Columbus, OH	1	30
		Shearon Harris, NC	199	1101632.6

State	Origination	Destination	Number of Shipments	Kg of Spent Fuel Shipped
	GE Wilmington, NC	Shearon Harris, NC	2	19
	McGuire, NC	Babcock & Wilcox, VA	1	5
		Dundalk, MD**	1	13.1
	Southport, NC*	Savannah River Site, SC	1	22
	Sunny Point, NC*	Savannah River Site, SC	1	14
ND	Pembina, ND*	Idaho National Engineering Lab	1	3
NE	Cooper, NE	GE Morris, IL	30	194546
	Fort Calhoun, NE	Battelle-Columbus, OH	1	1
	Veteran Admin, NE	Denver Federal Center	2	827
NJ	Hope Creek, NJ	GE Vallecitos, CA	1	17
	Oyster Creek, NJ	Battelle-Columbus, OH	1	33
NY	Alexandria Bay, NY*	Savannah River Site, SC	10	227
	Buffalo, NY*	Idaho National Engineering Lab	2	247
		Savannah River Site, SC	2	247
	Cintichem, NY	Idaho National Engineering Lab	3	10
		Savannah River Site, SC	14	41
	Cornell University, NY	Idaho National Engineering Lab	1	497
	Erie, NY*	Savannah River Site, SC	1	5.6
	Fort Erie, NY*	Savannah River Site, SC	1	3
	Ginna, NY	Dundalk, MD	1	4
	Nuclear Fuel Service, West Valley, NY	Battelle-Columbus, OH	8	2977
		Dresden Station, IL	31	20447
		Ginna, NY	73	32300
		Oyster Creek, NJ	33	42950
		Point Beach, WI	114	48450
	Ogdensburg, NY*	Savannah River Site, SC	14	35
OH	Battelle-Columbus, OH	Calvert Cliffs, MD	1	72
		GE Morris, IL	2	791
		GE Vallecitos, CA	2	28
		Ginna, NY	5	2632
		Savannah River Site, SC	1	1
		Zion, IL	2	879

State	Origination	Destination	Number of Shipments	Kg of Spent Fuel Shipped
OR	Portland, OR*	Idaho National Engineering Lab	28	139
PA	Limerick Station, PA	GE Vallecitos, CA	2	44
	Three Mile Island, PA	GE Vallecitos, CA	1	12.8
RI	Rhode Island AEC, RI	Savannah River Site, SC	35	16
SC	Charleston, SC*	Idaho National Engineering Lab	5	2855
		Savannah River Site, SC	149	18359.87
	Oconee, SC	Babcock & Wilcox, VA	6	972
		Buffalo, NY**	2	23
		McGuire, VA	138	140094
	Robinson, SC	Brunswick, NC	18	49725
		GE Vallecitos, CA	1	358
		Shearon Harris, NC	49	345666.5
	Savannah River Site, SC	Idaho National Engineering Lab	9	2714
	Virgil Summer, SC*	Alexandria Bay, NY	2	35
TX	Texas A&M University	Idaho National Engineering Lab	2	481
VA	Babcock & Wilcox, VA	GE Vallecitos, CA	2	41
		Oconee, SC	3	558
		Quad Cities, IL	2	499
	Newport News, VA*	Savannah River Site, SC	4	24
	Norfolk Intl Terminal, VA	Savannah River Site, SC	1	7
	North Anna Power Station, VA	Portsmouth, VA**	1	528
	Portsmouth, VA*	GE Vallecitos, CA	1	9
		Idaho National Engineering Lab	39	200
		Savannah River Site, SC	169	1057
	Surry, VA	Battelle-Columbus, OH	1	20
	Univ of Virginia, VA	Savannah River Site, SC	10	15
VT	Derby Lane, VT*	Savannah River Site, SC	8	20

Table 3-2 Summary Data for 1979 - 2007 Spent Fuel Shipment Information

Year	Number of Shipments		Kilograms Spent Fuel Shipped (Thousand)		Shipment Miles (Thousand)		Kilogram-Miles (Million)	
	Highway	Railway	Highway	Railway	Highway	Railway	Highway	Railway
1979	16	11	0.1	30.2	8.0	2.3	0.1	6.2
1980	130	5	10.1	13.6	115.9	1.0	17.2	2.8
1981	81	2	7.9	6.0	38.5	0.4	1.7	1.2
1982	124	0	7.1	0.0	106.8	0.0	1.8	0.0
1983	117	0	36.6	0.0	83.6	0.0	12.7	0.0
1984	245	3	84.5	23.8	181.3	1.6	51.4	12.7
1985	135	18	74.0	119.4	70.9	8.7	28.3	57.8
1986	105	15	40.4	97.5	47.8	8.7	8.8	56.3
1987	107	15	82.3	101.4	41.8	8.4	14.8	56.5
1988	25	7	12.8	41.8	11.4	4.3	2.4	25.7
1989	16	6	0.1	30.8	16.7	1.7	0.1	8.7
1990	5	8	(0.03)*	70.5	1.5	1.6	(0.02)*	12.7
1991	9	10	0.1	101.1	9.6	1.5	0.1	15.0
1992	21	6	0.1	61.2	15.7	0.8	0.1	8.1
1993	16	12	0.1	113.8	23.2	2.3	0.3	21.9
1994	7	10	(0.02)*	84.0	6.6	2.2	(0.01)*	17.4
1995	9	10	(0.07)*	83.7	12.6	2.2	0.1	18.4
1996	4	9	(0.02)*	77.9	2.8	1.5	(0.01)*	13.1
1997	9	6	(0.08)*	39.9	12.7	0.9	0.1	6.0
1998	11	15	0.6	105.3	11.6	3.7	0.6	22.3
1999	10	13	3.6	90.1	14.5	2.9	7.9	19.0
2000	15	5	2.8	42.0	18.5	0.9	2.3	5.9
2001	12	9	2.3	61.1	18.0	2.1	4.0	13.2
2002	8	18	1.7	163.0	7.2	3.4	1.0	30.3
2003	18	15	2.6	141.4	22.2	3.0	3.2	27.0
2004	9	16	2.6	163.0	9.6	3.1	3.3	32.9
2005	7	9	0.6	91.0	6.9	1.7	0.8	18.7
2006	6	8	0.9	64.4	7.1	1.6	1.5	13.2
2007	8	8	3.0	75.3	6.7	1.6	1.6	15.5
TOTAL	1285	269	377	2093	930	74	166	538

*Entries in parenthesis rounded to nearest hundredth. All others rounded to nearest tenth.

Table 3-3 Number of Domestic and International Spent Fuel Shipments from 1979 - 2007

Year	Domestic		International		
	Highway	Railway	Export	Import	Transshipments
1979	2	11	0	14	0
1980	73	5	2	55	0
1981	30	2	3	48	0
1982	80	0	1	43	0
1983	92	0	2	23	0
1984	209	3	2	34	0
1985	114	18	0	21	0
1986	88	15	0	17	0
1987	85	15	3	19	0
1988	10	7	0	15	0
1989	11	6	1	4	0
1990	0	8	3	0	2
1991	4	10	4	0	1
1992	20	6	0	0	1
1993	14	12	1	0	1
1994	6	9	1	1	0
1995	7	9	1	2	0
1996	3	8	0	2	0
1997	7	4	1	3	0
1998	11	11	0	4	0
1999	8	9	0	6	0
2000	10	4	0	6	0
2001	9	6	1	5	0
2002	6	16	2	2	0
2003	15	14	1	3	0
2004	7	14	0	4	0
2005	6	7	0	3	0
2006	5	7	0	2	0
2007	6	7	0	3	0

Total Number of Shipments - 1553

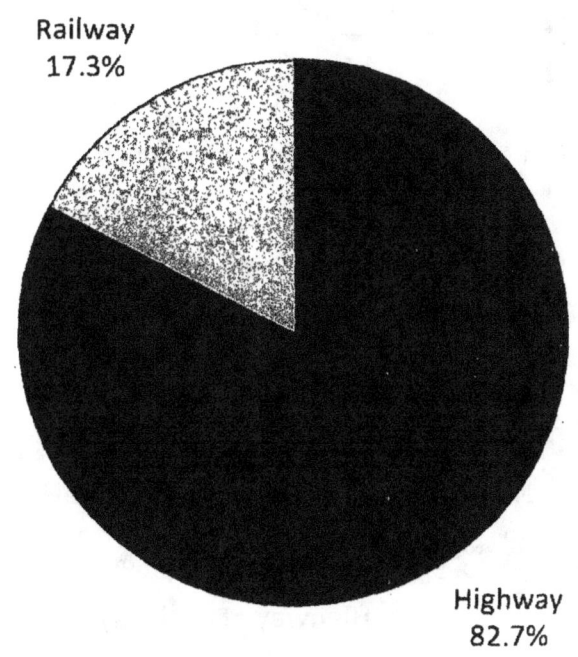

Railway
17.3%

Highway
82.7%

Figure 3-2 Number of Spent Fuel Shipments by Mode from 1979 - 2007

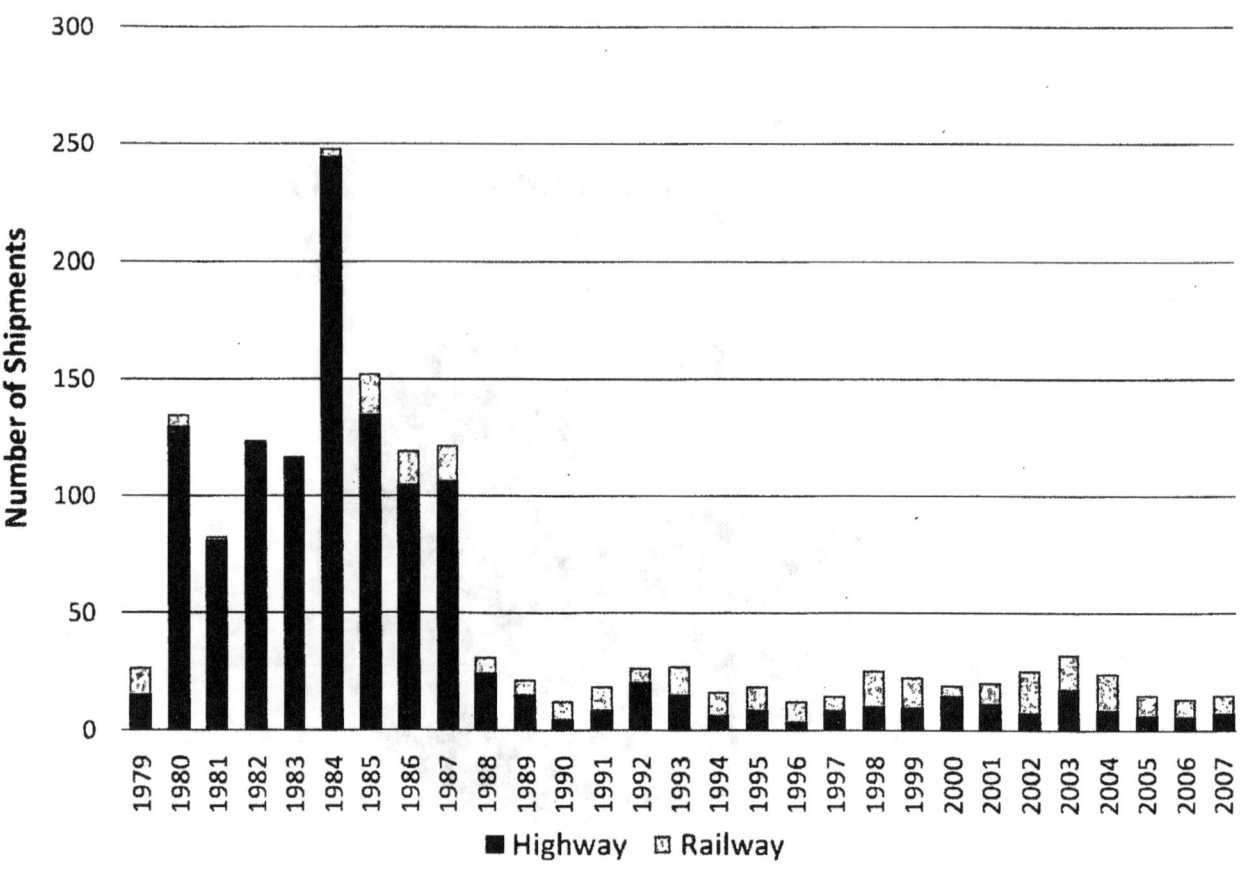

Figure 3-3 Number of Spent Fuel Shipments by Year from 1979 - 2007

Total Kilograms of Spent Fuel Shipped - 2,470,000

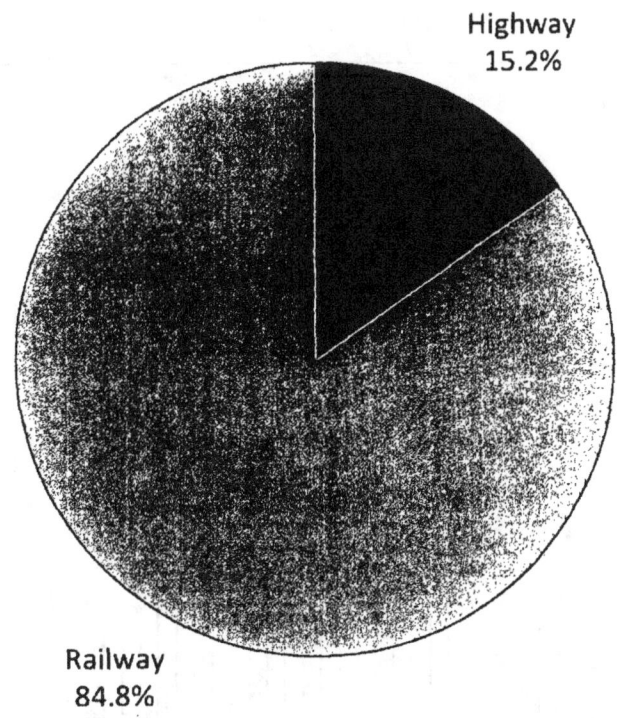

Highway
15.2%

Railway
84.8%

Figure 3-4 Quantity of Spent Fuel Shipped by Mode from 1979 - 2007

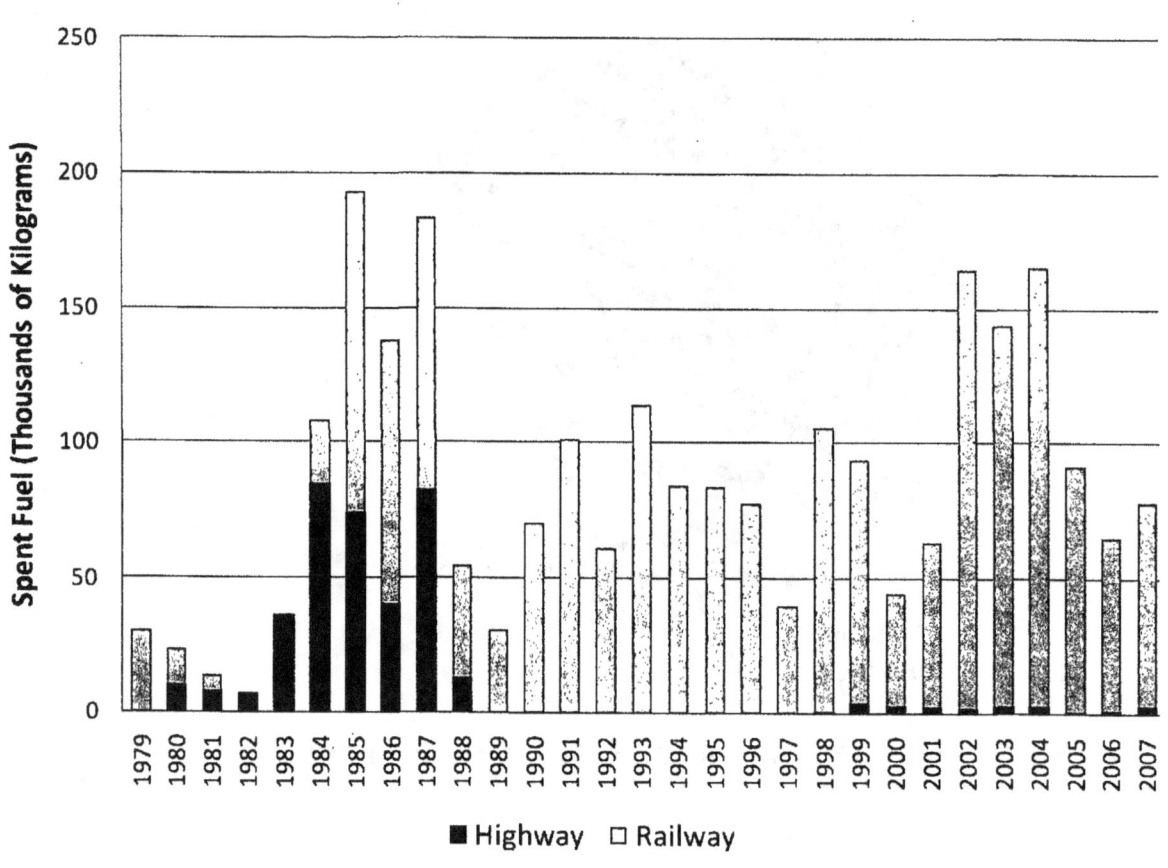

Figure 3-5 Quantity of Spent Fuel Shipped by Year from 1979 - 2007

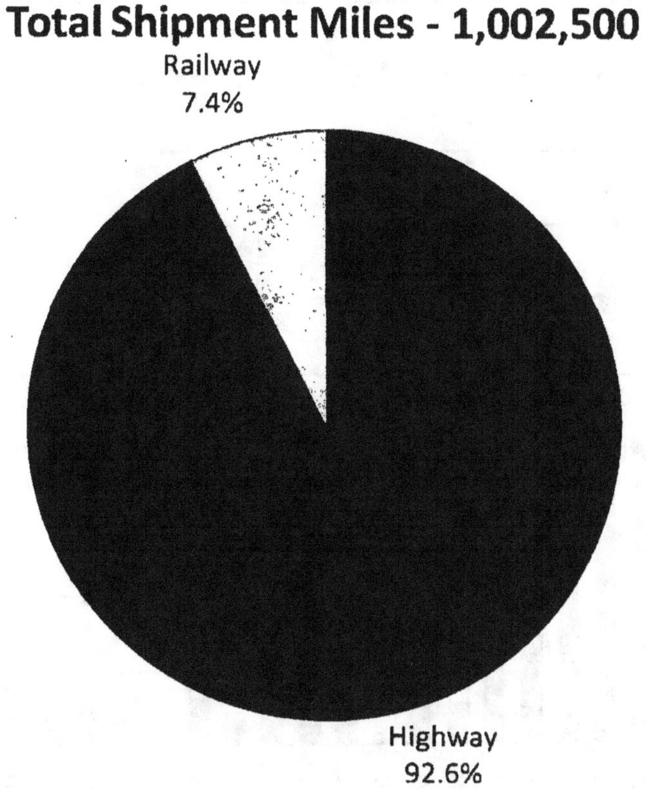

Total Shipment Miles - 1,002,500

Railway
7.4%

Highway
92.6%

Figure 3-6 Spent Fuel Shipment Miles by Mode from 1979 - 2007

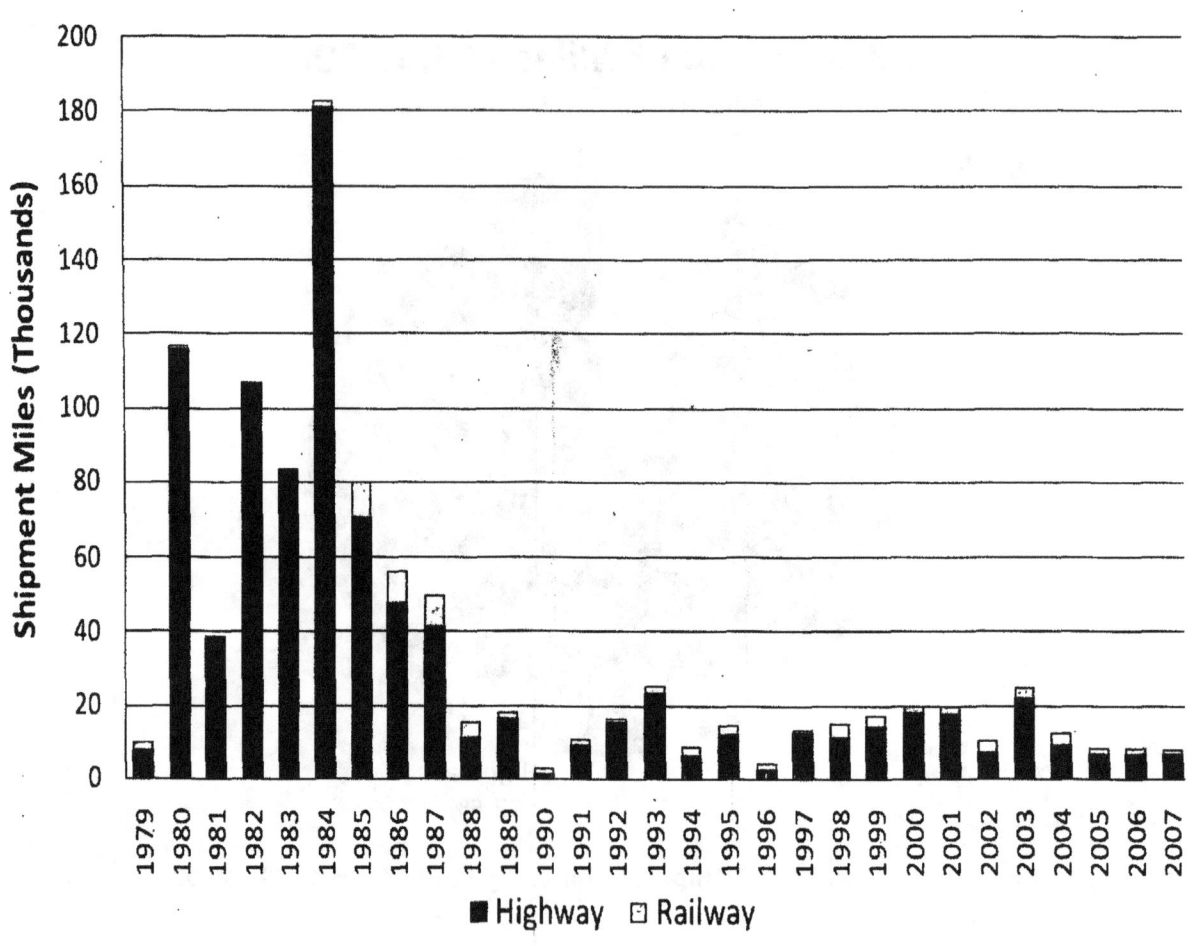

Figure 3-7 Spent Fuel Shipment Miles by Year from 1979 - 2007

Total Kilogram-Miles - 704,540,000

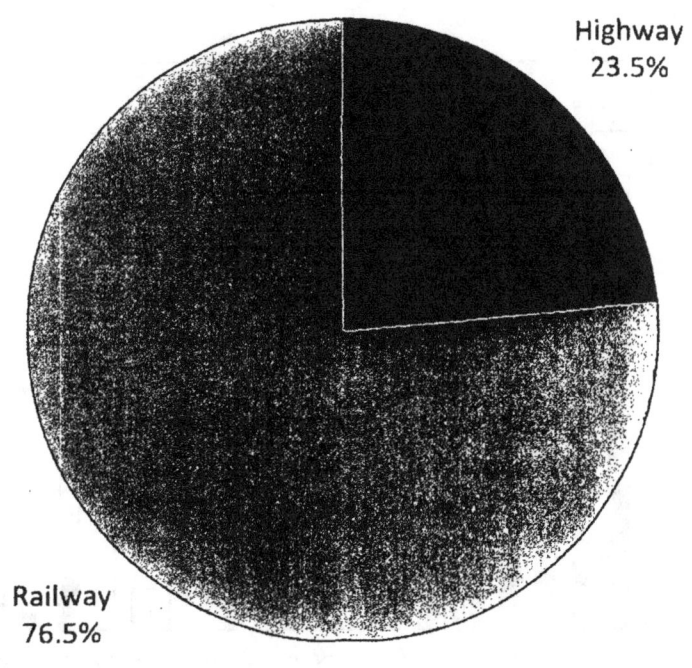

Highway
23.5%

Railway
76.5%

Figure 3-8 Spent Fuel Shipment Kilogram-Miles by Mode from 1979 - 2007

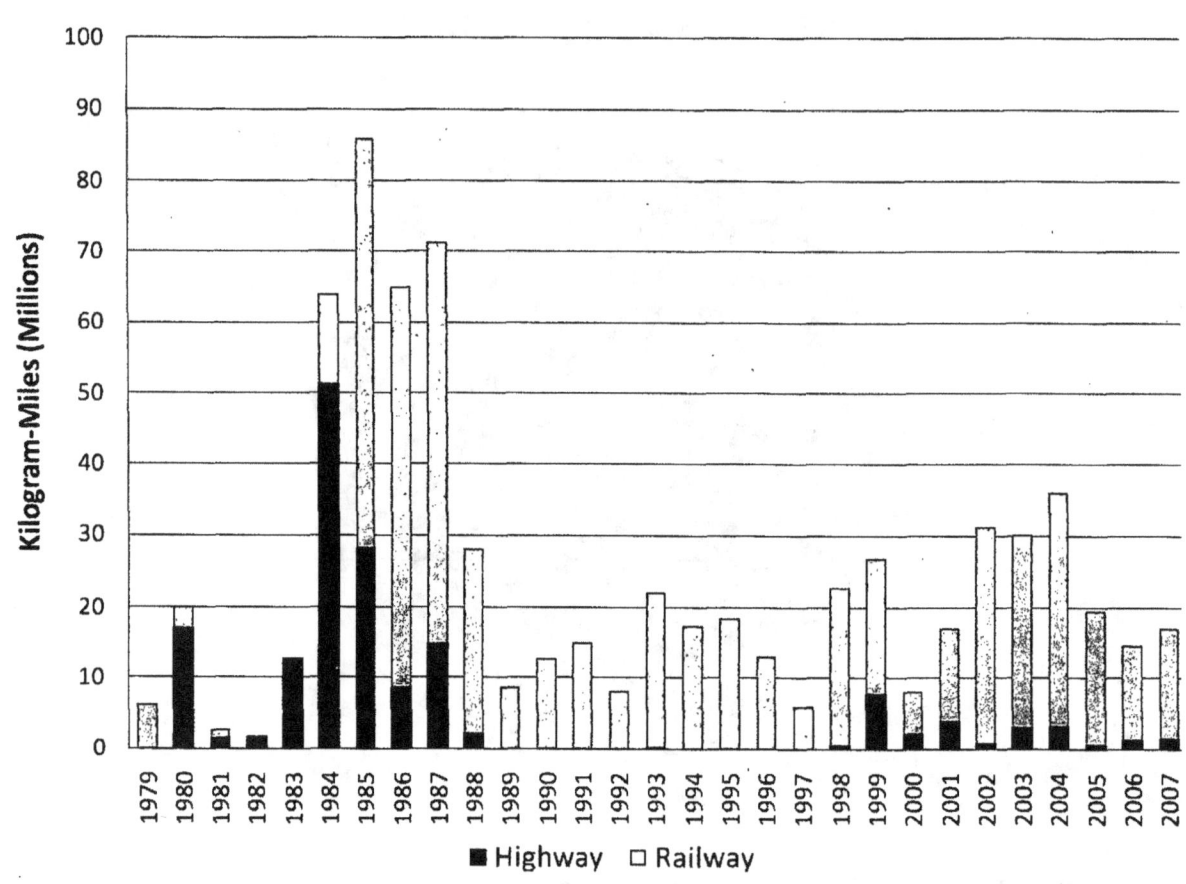

Figure 3-9 Spent Fuel Shipment Kilogram-Miles by Year from 1979 - 2007

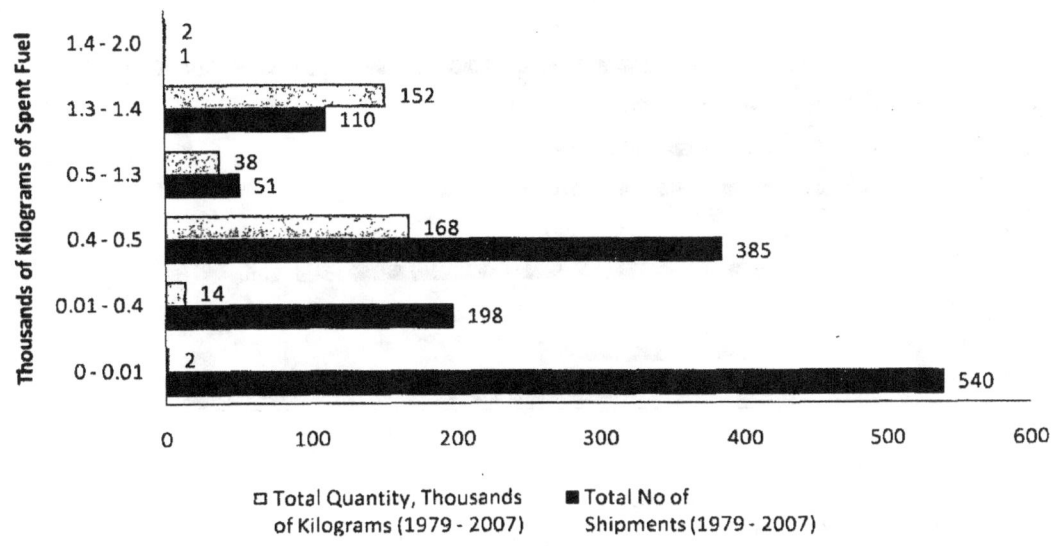

Highway Shipment Quantity
1979 - 2007

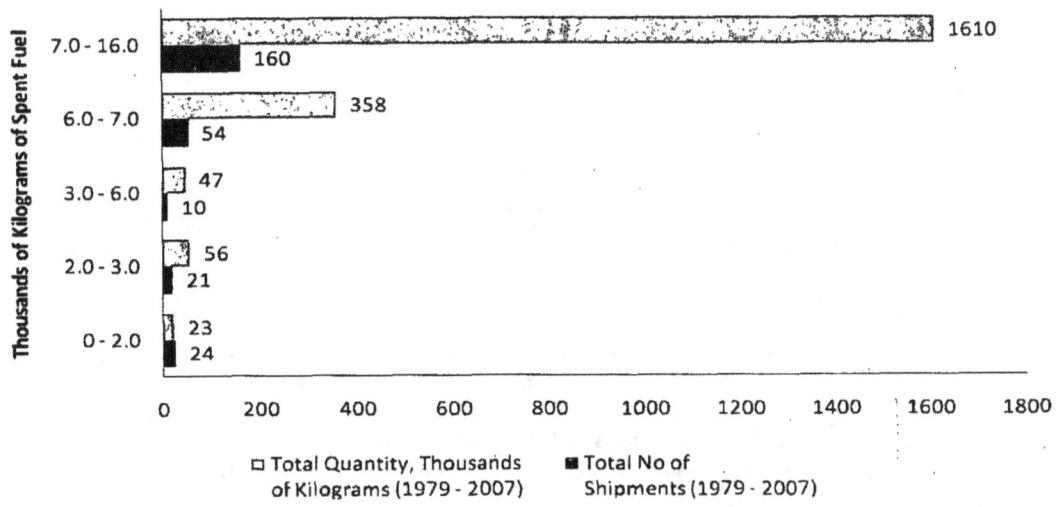

Railway Shipment Quantity
1979 - 2007

Figure 3-10 Number and Total Quantity of Spent Fuel Shipments by Shipment Quantity Range from 1979 - 2007

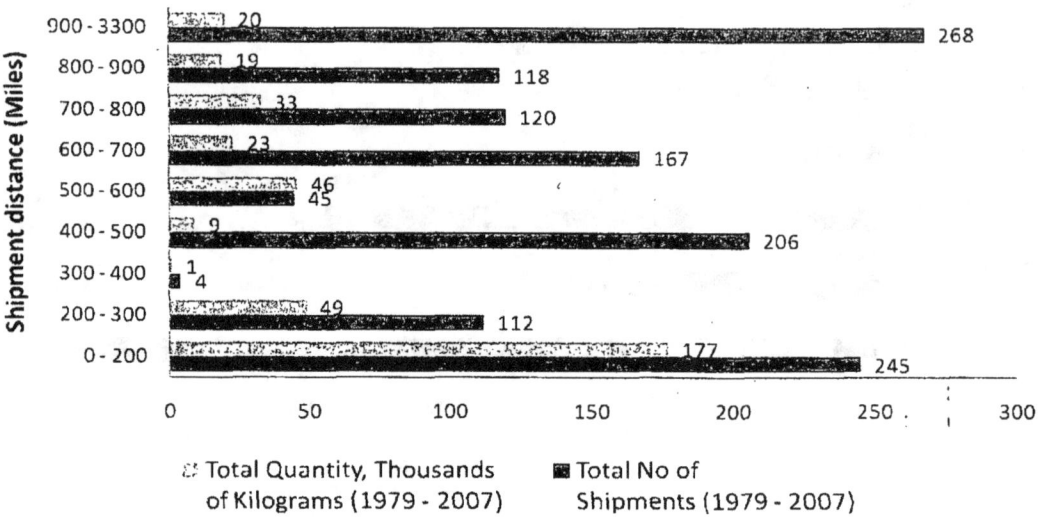

Highway Shipment Distance 1979 - 2007

☐ Total Quantity, Thousands of Kilograms (1979 - 2007)   ■ Total No of Shipments (1979 - 2007)

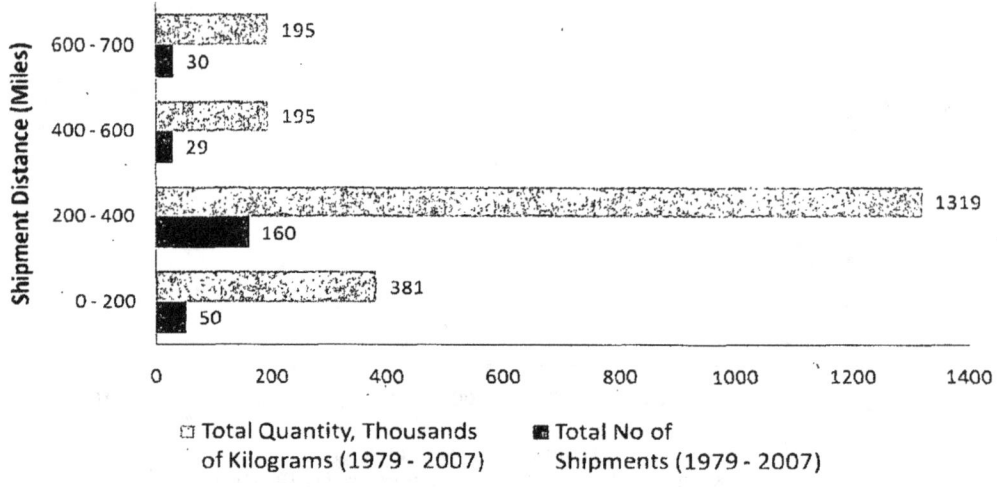

Railway Shipment Distance 1979 - 2007

☐ Total Quantity, Thousands of Kilograms (1979 - 2007)   ■ Total No of Shipments (1979 - 2007)

Figure 3-11 Number and Total Quantity of Spent Fuel Shipments by Shipment Distance Range from 1979 - 2007

Table 4-1 Highway and Railway Spent Fuel Shipment Routes Used from 1998 - 2007

Origin State	Shipment	NRC Route Number	Transport Mode	Shipment Years
Alabama	Browns Ferry Nuclear Station to GE Vallecitos Nuclear Center	209	Highway	2003
California	GE Vallecitos Nuclear Center to Argonne National Laboratory	190	Highway	2000, 2001
	General Atomic to Bechtel BWXT, INEEL	207	Highway	2003
Florida	University of Florida to Savannah River Site	222	Highway	2006, 2007
Illinois	La Salle County Station to Newport News	208	Highway	2003
	University of Illinois to INEEL	214	Highway	2004
	University of Illinois to University of Texas, Austin	213	Highway	2004
Indiana	Purdue University Training Reactor to Savannah River Site	225	Highway	2007
Iowa	Duane Arnold Energy Center to GE Vallecitos Nuclear Center	163	Highway	1998, 2008
Maryland	Dundalk Marine Terminal to GE Vallecitos Nuclear Center	180	Highway	2000
	National Institute of Standards and Technology to Savannah River Site	187	Highway	1999, 2003
Massachusetts	Massachusetts Institute of Technology to Savannah River Site	166	Highway	1998 thru 2002
	University of Massachusetts, Lowell to Savannah River Site	215	Highway	2004
	Massachusetts Institute of Technology to Savannah River Site	217	Highway	2005 thru 2007
Michigan	University of Michigan to Savannah River Site	196	Highway	1999, 2000, 2003
Missouri	University of Missouri, Columbia to Savannah River Site	182	Highway	1998 thru 2004
	University of Missouri, Columbia to Savannah River Site	182B	Highway	2005 thru 2007
Nebraska	Veteran Administration to U.S. Geological Survey, Denver Federal Center	206	Highway	2002
New York	Cornell University to Bechtel BWXT, INEEL	212	Highway	2003
	McMaster University to Savannah River Site	198	Highway	2000

Origin State	Shipment	NRC Route Number	Transport Mode	Shipment Years
New York	University of Toronto to Savannah River Site	198	Highway	2000
	University of NY, Buffalo, NY to Idaho National Laboratory, Scoville, ID	216	Highway	2005
North Carolina	Brunswick Nuclear Plant to Harris Nuclear Plant	130	Railway	1998, 1999, 2001 thru 2007
Ohio	Battelle .West Jefferson Site to Savannah River Site	211	Highway	2003
Pennsylvania	Limerick Generating Station to GE Vallecitos Nuclear Center	197	Highway	1999, 2003
South Carolina	H.B. Robinson Steam Electric Plant to Harris Nuclear Plant	135	Railway	2000, 2002 thru 2004
	H.B. Robinson Steam Electric Plant to GE Vallecitos Nuclear Center	200	Highway	2001
	Charleston to Savannah River Site	185	Railway	1999, 2001, 2002, 2004 thru 2007
	Charleston to Savannah River Site	192	Highway	2000, 2001, 2004, 2007
	Charleston to Savannah River Site	201A	Railway	1998 thru 2003
	Charleston to Savannah River Site	210	Highway	2003, 2005
	Charleston to INEEL	192 & 195	Highway	1999
	Savannah River Site to INEEL	195	Highway	2000, 2001, 2003
	Savannah River Site to INEEL	202	Highway	2004, 2006
	Oconee Nuclear Site to AECL Chalk River	203	Highway	2001, 2002
Texas	Texas A&M University to INEEL	221	Highway	2006, 2007
Virginia	North Anna Power Station to Studsvik Nuclear	204A	Highway	2002

NRC FORM 335 (9-2004) NRCMD 3.7	U.S. NUCLEAR REGULATORY COMMISSION	1. REPORT NUMBER (Assigned by NRC, Add Vol., Supp., Rev., and Addendum Numbers, if any.)
	BIBLIOGRAPHIC DATA SHEET *(See instructions on the reverse)*	NUREG-0725, Rev. 15

2. TITLE AND SUBTITLE	3. DATE REPORT PUBLISHED	
Public Information Circular For Shipments of Irradiated Reactor Fuel	MONTH	YEAR
	05	2010
	4. FIN OR GRANT NUMBER	
	N4104	

5. AUTHOR(S)	6. TYPE OF REPORT
A. G. Garrett, S. L. Garrett, and R. G. Ostler	Annual
	7. PERIOD COVERED *(Inclusive Dates)*
	1979-2007

8. PERFORMING ORGANIZATION - NAME AND ADDRESS *(If NRC, provide Division, Office or Region, U.S. Nuclear Regulatory Commission, and mailing address; if contractor, provide name and mailing address.)*

Pacific Northwest National Laboratory
P.O. Box 999
Richland, WA 99352

9. SPONSORING ORGANIZATION - NAME AND ADDRESS *(If NRC, type "Same as above"; if contractor, provide NRC Division, Office or Region, U.S. Nuclear Regulatory Commission, and mailing address.)*

Office of Nuclear Security and Incident Response
U.S. Nuclear Regulatory Commission
Washington, DC 20555-0001

10. SUPPLEMENTARY NOTES

Statistics for Highway and Railway Spent Fuel Shipments from 1979-2007

11. ABSTRACT *(200 words or less)*

This report provides information on the shipment of irradiated reactor fuel (spent fuel) subject to regulation by the U.S. Nuclear Regulatory Commission (NRC). It briefly describes spent fuel shipment safety and safeguards requirements of general interest, summarizes data for highway and railway shipments from 1979 - 2007, and lists, by State, recent highway and railway shipment routes. This circular does not include Department of Defense and Department of Energy spent fuel shipments.

The enclosed route information reflects specific NRC approvals that the agency has granted in response to requests for shipments of spent fuel. This publication does not constitute authority for carriers or other persons to use the routes to ship spent fuel, other categories of nuclear waste, or other materials.

12. KEY WORDS/DESCRIPTORS *(List words or phrases that will assist researchers in locating the report.)*	13. AVAILABILITY STATEMENT
Shipments of Irradiated Reactor Fuel, Shipments of Spent Fuel, Regulatory Requirements for Spent Fuel Shipments, Statistics for Highway and Railway Shipments of Spent Fuel ,Packaging of Spent Fuel, Spent Fuel Shipment Routes, Spent Fuel Shipment Information, Physical Protection Requirements of Irradiated Reactor Fuel in Transit.	unlimited
	14. SECURITY CLASSIFICATION
	(This Page)
	unclassified
	(This Report)
	unclassified
	15. NUMBER OF PAGES
	16. PRICE

NRC FORM 335 (9-2004) PRINTED ON RECYCLED PAPER

UNITED STATES
NUCLEAR REGULATORY COMMISSION
WASHINGTON, DC 20555-0001

OFFICIAL BUSINESS